SPACE FIRSTS™

COLUMBIA
The First Space Shuttle

Heather Feldman

The Rosen Publishing Group's
PowerKids Press™
New York

For Joshua and Jeremy Hochman—may all your dreams come true. YLA.

Published in 2003 by The Rosen Publishing Group, Inc.
29 East 21st Street, New York, NY 10010

First Edition

Editor: Nancy MacDonell Smith
Book Design: Mike Donnellan

Photo Credits: Cover, pp. 4, 7, 8, 11, 15, 16 © Photri Microstock, Inc.; p. 12 © James L. Amos/CORBIS; pp. 19, 20 © NASA/Roger Ressmeyer/CORBIS.

Feldman, Heather.
Columbia : the first space shuttle / Heather Feldman.— 1st ed.
     p. cm. — (Space firsts)
Includes bibliographical references and index.
Summary: Chronicles the building and lift-off of Columbia, the first space shuttle.
ISBN 0-8239-6247-4 (library binding)
1. Space shuttles—Juvenile literature. 2. Columbia (Spacecraft)—Juvenile literature. [1. Space shuttles. 2. Columbia (Spacecraft)] I. Title.
TL795.515 .F45 2003
629.44'1—dc21

                          2001007785

Manufactured in the United States of America

Contents

The First Space Shuttle

After the first astronauts landed on the Moon in 1969, scientists began thinking about setting up a **permanent** space station to gather information about space. A space station is a place where astronauts can live and work for long periods of time. Before scientists could build a space station, they needed a way to get astronauts to and from space. This is why **engineers** at the National Aeronautics and Space Administration (NASA) developed the first **space shuttle**. A space shuttle looks a little like an airplane and a little like a rocket. The first space shuttle scientists built never actually reached space. Scientists worked very hard to perfect the space shuttle. Their hard work paid off, and their dream of a space shuttle in space was about to come true.

Scientists built Enterprise in 1976. Enterprise looked similar to later shuttles, but it never reached space.

Space travel changed forever in 1981. That year the space shuttle *Columbia* was **launched**. This was the first time a spacecraft went to space, returned to Earth, and was able to go back to space again. Until this time, only rockets had been used successfully in space travel. Rockets can only be used once for such a flight. This was very expensive for the space program.

A space shuttle is different. It is launched like a rocket, but it returns to Earth like an airplane. A space shuttle has three main parts. After it returns to Earth, two of the shuttle's three main parts can be used again and again.

Since *Columbia*, several other space shuttles have been built. All were designed and built to look the same.

The front end of Columbia looks very much like the front end of an airplane. Columbia was built at the Kennedy Space Center, in Florida.

fuel tank

booster
rockets

orbiter

The Parts of a Space Shuttle

There are three main parts to each shuttle. The first part is the **orbiter**. The orbiter is the part of the shuttle that carries the crew and the **cargo**. It looks like an airplane. This is the part of the shuttle that actually makes it to space, **orbits** Earth, and returns home. The second part of the shuttle is the fuel tank. It is taller than a 15-story building and carries liquid fuel for the orbiter's engines. The fuel tank is 154 feet (47 m) long. After the launch the fuel is quickly emptied from the giant fuel tank. Then the fuel tank falls off and breaks up in the **atmosphere**. The third part, the **booster rockets**, provide the power to launch the orbiter toward space. The rockets are covered with a special material that keeps them from breaking up.

The fuel tank is the largest part of a space shuttle. The fuel tank does not get reused. The orbiter returns to Earth with the astronauts. The booster rockets drop into the ocean and float so they can be picked up and used again.

Inside a Space Shuttle

Inside the space shuttle, the astronauts have a lot of work to do. There is not a whole lot of room inside, but the astronauts are able to stay comfortable. On *Columbia*'s first voyage, there were only two astronauts. Space shuttles are designed to hold up to 10 astronauts.

The cabin has two levels. The pilot flies the space shuttle from the flight deck. The crew stays on the lower level, called the middeck. The bathroom, sleeping bunks, kitchen, and lockers are all on the middeck. The middeck can be a very crowded place! There is also an area called the **payload bay**. This is where the astronauts keep the cargo. Cargo is what the astronauts bring on board to do their work in space. Cargo can be equipment to do experiments, or parts to fix a broken **satellite**.

This is the flight deck of the Columbia. *It is located on the upper level. The control panel on the flight deck has more than 2,000 switches and dials!* Inset: *Equipment for experiments is kept in the payload bay of* Columbia.

Hot Tiles

When a spacecraft returns to Earth, it travels through space and reenters Earth's atmosphere. Reentering the atmosphere causes **intense** heat. Temperatures can reach 2,500°F (1,371°C). This can severely damage a spacecraft. Before the space shuttle was designed, rockets used for space flight would reenter the atmosphere and burn up. Astronauts had to **eject** quickly from a rocket before it got too hot. They would then **parachute** down to Earth. Scientists needed to design a space shuttle that could withstand the intense heat as it flew through this layer of air that surrounds Earth. Scientists figured out a way to protect a shuttle from this heat. They covered the space shuttle with a layer of special tiles. These **heat-resistant** tiles cover the entire surface of a space shuttle.

More than 31,000 tiles covered the outside of Columbia. Each tile was carefully tested and attached with a very powerful glue.

Columbia's Mission

By the spring of 1981, *Columbia* was ready for her **mission**. The astronauts on board were John W. Young, the commander, and Robert L. Crippen, the pilot. *Columbia's* mission had several goals. The first goal was to complete simple working **procedures** of the shuttle system. The second goal was to make a safe trip into space. The third goal was to make a safe landing back on Earth. The crew would be carrying a package that contained tools used to record how the orbiter performed on the flight. The astronauts had to test the shuttle's equipment. One important exercise was to open and close the large payload-bay doors. This would be done using computers. The mission was clear, the astronauts were ready, and the journey was about to begin.

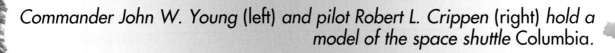

Commander John W. Young (left) *and pilot Robert L. Crippen* (right) *hold a model of the space shuttle* Columbia.

Liftoff!

At 7:00 A.M. on April 12, 1981, the space shuttle *Columbia* was launched at the Kennedy Space Center, in Florida. As the shuttle took off, the ground shook and the engines roared. During liftoff, the shuttle rode on the back of a huge rocket. The rocket engines swallowed up liquid fuel. The space shuttle was powerfully thrust off the launchpad. *Columbia* reached for the sky, followed by a trail of smoke and fire. Two minutes after liftoff, the booster rockets had finished their work. They were cut loose and dropped into the Atlantic Ocean. About eight minutes after liftoff, the big fuel tank was empty and was also detached from the shuttle, most of it breaking up in the atmosphere. Minutes later, *Columbia* was in space! The dream of a space shuttle was a reality.

About 10 minutes after this photo was taken, Columbia *traveled at speeds of 17,400 miles per hour (28,003 km/h)!*

17

Columbia in Space

The astronauts on board were ready for this mission. They had trained long and hard. Now the astronauts needed to do their work. They had to complete simple tasks and test all the space shuttle's equipment. They tested the payload-bay doors, which opened exactly as planned. During the flight, televised pictures of the astronauts at work or on breaks were sent back to Earth.

The televised pictures also showed that some of the tiles used to protect *Columbia* from the intense heat were missing. Many others were damaged. This probably happened during liftoff, from all the shaking the shuttle does. Even so the mission was able to continue as planned. For two days, *Columbia* circled Earth successfully.

Columbia traveled in space for 54 hours and 21 minutes. It orbited Earth 36 times. This flight was the first space mission for the pilot, Robert L. Crippen.

Return to Earth

On April 14, 1981, *Columbia* completed its final orbit around Earth and began the journey home. Computers helped to guide *Columbia* back to Earth for a while, then Commander Young took over the controls. For about 21 minutes, as *Columbia* traveled back into Earth's atmosphere, the shuttle was cut off from all communications with mission control. Then, suddenly, Commander Young was heard saying, "Hello Houston. *Columbia* here!" Everyone was very excited and waited anxiously for *Columbia* to return to Earth. More than 350,000 people came to watch *Columbia* land at Edwards Air Force Base in California, and millions more watched on television. At 10:21 A.M., they watched as *Columbia* came home, making a perfect landing.

Columbia *makes a perfect landing.* Columbia's *first flight was so successful that the space shuttle went on to fly the next four shuttle missions.*

Other Space Shuttles

The next space shuttle was *Challenger*, which was built in 1982. Then came *Discovery* in 1983, which was lighter and stronger. *Discovery* helped to send the **Hubble Space Telescope** into orbit. This telescope has provided an unbelievably clear view into the universe. *Atlantis* joined the fleet of space shuttles in 1985. In 1986, *Challenger* was about to go on its tenth flight. Then something went terribly wrong. *Challenger* exploded just after takeoff. It took a long time for NASA to send another shuttle into space, but it did, on September 29, 1988. This flight was successful. Since then, shuttles such as *Endeavor*, which arrived in 1991, have continued to help fix and launch satellites. Shuttles have also been important in the development of space stations. Shuttles are a very important part of space exploration. They help us gain knowledge about and an understanding of our universe.

Glossary

atmosphere (AT-muh-sfeer) The layer of gases that surrounds an object in space. On Earth, this layer is air.

booster rockets (BOO-ster RAH-kits) Rockets used as main sources of thrust in takeoff.

cargo (KAR-goh) Goods carried by a spacecraft, an airplane, or a ship.

eject (ee-JEKT) To drive or throw out from within.

engineers (en-jih-NEERZ) People who use science to design and build things.

heat-resistant (HEET ree-ZIS-tent) Able to survive very high temperatures.

Hubble Space Telescope (HUH-bul SPAYS TEL-uh-skohp) A telescope launched into space in April 1990 that has sent back many images of Uranus and other planets.

intense (in-TENTS) Very strong.

launched (LAWNCHD) When a spacecraft was pushed into the air.

mission (MIH-shun) Being sent somewhere to do special work.

orbiter (OR-bih-ter) The main part of a shuttle that carries the crew and cargo.

orbits (OR-bits) Travels in a circular path around an object in space.

parachute (PAR-uh-shoot) To descend at a safe rate of speed to the ground by using a large, umbrellalike device of fabric that opens in midair.

payload bay (PAY-lohd BAY) The inside part of a shuttle where astronauts keep their cargo.

permanent (PUR-muh-nint) Lasting forever or for an indefinitely long time.

procedures (pro-SEE-jerz) A series of steps or rules to follow.

satellite (SA-til-eyet) A human-made or natural object that orbits another body.

space shuttle (SPAYS SHUH-tul) A reusable spacecraft designed to travel to and from space carrying people and cargo.

Index

Web Sites

Due to the changing nature of Internet links, PowerKids Press has developed an online list of Web sites related to the subject of this book. This site is updated regularly. Please use this link to access the list:
www.powerkidslinks.com/sf/columb/